Microbiology: Understanding Viruses

Why I Wrote This Book

Viruses are tiny, microscopic particles - the smallest known viruses are so small that two hundred thousand of them could fit onto a pin head.

They are the most abundant type of microbe, more numerous than bacteria or fungi, archaea or protista.

The first virus ever discovered was the tobacco mosaic in 1898. Thousands of different types of viruses are now known although it is thought that there are millions more types that are as yet, undiscovered.

Viruses are important to us for a number of reasons.

On the negative side: viruses cause diseases such as AIDS, Ebola, rabies, polio, yellow and dengue fever as well as the usually less serious though more common; flu and the cold. Genetically engineered viruses might have the capacity to cause death and devastation to large populations. This has led to the concern that viruses could be used in biological warfare.

On the other hand viruses are very useful in research, particularly in the fields of molecular biology, genetic engineering and medicine. Viruses can be used in eliminating harmful crop pests such as insects. Viruses are also useful in the preparation of vaccines.

In this book I aim to give you a summary of the main aspects of viruses: their structure, uses, health impact and evolutionary origin.

The main points are made in a clear and easily understood question and answer format.

The book starts with a few interesting facts to whet your appetite and then explains the main scientific terms so you know them before you read the main aspects.

We are all affected by viruses regarding our health alone, on top of this they are a fascinating area of biology to study. The HIV, swine flu and Ebola epidemics have only served to illustrate the impact that viruses can have on human health.

I hope you enjoy reading this book and it helps you in your biology studies. That is why I wrote it.

A Few Fascinating Facts

1. Flu vaccines change every year, as there are a number of different flu viruses, and these viruses mutate (change) approximately every year.

2. There are a million virus particles per millilitre of seawater - so around the whole world the sea contains many trillions of viruses.

3. Walter Reed discovered the first human disease-causing virus; the yellow fever virus, in 1901.

4. Some viruses are useful – viruses called bacteriophages (means 'bacteria eater') kill bacteria and are used to protect people against harmful bacteria in food.

5. Viral infections cannot be cured with antibiotics – most common viral infections, like a cold or flu, just get better without medical help, thanks to the action of our immune system.

6. Viruses are passed on by sneezing and coughing, or by touching someone who has a viral infection – this is why it is important to cough or sneeze into tissues and wash hands regularly when infected with a cold or other viral infection. Sometimes viruses can be caught by touching a surface with the viruses on it.

7. Over eight thousand deaths have been reported from Ebola since the start of the 2014 West Africa outbreak. There have been a few cases in the UK of medical staff returning from West Africa with Ebola disease.

Key Words Explained

RNA - Ribose Nucleic Acid - this is one of the forms of genetic material found in viruses.

mRNA - this is a form of RNA that copies a gene and takes this copy to the ribosome in the living cell where the particular protein coded for by this particular gene is then made.

DNA - Deoxyribose Nucleic Acid - this is one of the forms of genetic material found in viruses.

Capsid - the protein coat that envelopes viral genetic material is known as a capsid. It is composed of protein subunits called capsomeres. Capsids can have several shapes: polyhedral, rod or complex. Their function is to protect the viral genetic material from damage.

Retroviruses - are viruses with their genetic material as RNA. They replicate by formation of DNA by reverse transcription of their RNA.

HIV - Human Immunodeficiency Virus. This is the virus that causes the disease known as AIDS. Over thirty million people worldwide are believed to be infected with the HIV virus.

Ebola - a disease of humans and other primates (that can also be carried by fruit bats) that is caused by viruses belonging to the genus Ebolavirus. It is believed to have spread to humans originally through handling, skinning and eating infected fruit bats. The current outbreak is mainly confined to West Africa: Guinea; Liberia and Sierra Leone.

Protein - a large biological polymer molecule mainly made up of a string of amino acids joined together.

Enzyme - this is a globular (ball shaped) protein molecule which is a specific catalyst that speeds up a particular biochemical reaction.

Antibiotic - this is a drug that is given to combat bacterial infections; they have no effect against viruses.

Assembly - the construction of viral proteins within a living cell.

Bacteriophage - a virus that only infects bacterial cells.

Adsorption - the virus binding to the outside of the cell.

Penetration - the virus injecting its genetic material (either DNA or RNA) into the cell it is infecting.

Ribosome - an organelle found in living cells, where the sequence of bases in a molecule of mRNA are translated into a polypeptide chain of amino acids that makes up a protein.

Host Cell - this is a cell that is infected by a virus, which will replicate inside it.

Reverse transcriptase - this is an enzyme that allows the creation of DNA from RNA. It is most commonly found in retroviruses; such as HIV.

Antibiotic - this is a chemical that kills or slows the growth of bacterial cells without significantly affecting the human patient's cells. They are used against bacterial infections but have no effect on viral infections.

Gene Therapy - this is the introduction of new, functional genes into the cells of a patient suffering from a genetic disease; in order to replace the inherited, non-functioning genes the patient has.

Chapter One

Viruses - The Main Facts

What Are Viruses?

Viruses are the simplest and tiniest of microbes.

They are many times smaller than bacteria. Most viruses are about 0.1 micrometre across (that is one ten-thousandth of a millimetre).

They have a very simple structure; generally consisting of just a small collection of genetic material (DNA or RNA) encased in a protective protein coat.

They are not living things and they are not cells.

In fact, when they are outside of a host cell, viruses are inert, just mere microbial particles or crystals. Another way to look at them is as molecular syringes which are just moving their RNA or DNA between cells.

Viruses can only replicate within a living host cell. However, as they can replicate and have DNA and RNA which can mutate; they are capable of evolving.

Where Are Viruses Found?

Viruses are found just about everywhere: in every material and environment on Earth.

They do need to be able to get into living cells though; to infect the cells; in order to reproduce themselves. In fact, viruses have evolved to infect every form of life, from animal to plant and from fungi to bacteria.

What Kind Of Cells Do Viruses Infect?

A particular kind of virus tends to infect a particular type of cell. For example the HIV virus infects the T-helper white blood cells that form part of the human immune system.

On the other hand, some viruses may be able to infect more than one species; doing no harm in one of them, but having catastrophic consequences when it gets into a different creature. For example, the Hantavirus has almost no effect on deer mice which carry it, but if it infects people it can cause serious bleeding.

What Is The Structure of a Virus?

A virus is not a cell. It is not a living thing at all.

They are often described as particles, sometimes as crystals.

They are essentially a small piece of genetic material (DNA or RNA) inside a surrounding protein coat.

What Different Types Of Viruses Are There?

The shapes of viruses falls into a few different categories: rods, or filaments, or spherical (which actually take the form of twenty-sided polygons, called icosahedrons).

Helical

These are either rod-shaped or filamentous shaped (Ebola virus). They can be short and highly rigid (the tobacco mosaic virus) or long and very flexible.

Icosahedral

Most animal viruses are either the exact or the approximate shape of an icosahedron - a regular 3D shape with twenty identical equilateral triangular faces (rotavirus).

Prolate

This is an elongated icosahedron shape and is the common structure of the heads of bacteriophages.

Envelope

Some species of virus envelop themselves in a one of the host cell membranes: the outer membrane surrounding an infected host cell or internal membranes such as nuclear membrane. The virus then gains an outer lipid bilayer known as a viral envelope. The influenza virus and HIV are envelope viruses.

Complex

These viruses have a complex structure that is neither helical nor icosahedral, and may have protein tails or a complex outer wall. Some bacteriophages have a complex structure consisting of an icosahedral head bound to a helical tail in addition to a base plate with protruding protein tail fibres. This tail structure allows the virus to inject its genetic material into the host cell.

How Big Are Viruses?

All viruses are extremely small, although they do vary in size depending on their type.

Some are as small as fifteen nanometres in diameter (that is about fifteen millionths of a millimetre), although some types can be around ten to twenty times bigger.

What is the Genetic Material of Viruses?

Viruses may have their genetic material in the form of double-stranded DNA, double-stranded RNA, single-stranded DNA or single-stranded RNA; depending on the nature and function of the specific virus.

The viral genome can be made up of just a very small number of genes or a few hundred of genes; depending on the type of virus. The genetic material is usually organized in the form of a long molecule that may be straight or circular.

Chapter Two

How Do Viruses Replicate?

Why Do Viruses Need To Reproduce Inside Living Cells?

Viruses only exist to make more viruses.

They are not living things and therefore cannot replicate or express their genes without infecting and taking over a living cell, whether a plant, animal, fungus or bacterial cell. This is because a virus particle does not have the necessary components in order to reproduce. These components are actually found inside living cells.

So viruses need to take over the host cell's reproductive machinery in order to reproduce themselves.

Therefore viruses infect living cells and reproduce within them.

How Do Viruses Get Inside Living Cells?

Upon landing on an appropriate host cell, a virus gets its genetic material inside the cell either by tricking the host cell to pull the whole virus inside, like it would a nutrient molecule, or by fusing its viral coat with the host cell wall or membrane and releasing its genes inside.

Bacteriophages inject their genetic material into a host bacterial cell through their helical tail structure but their capsid remains outside of the cell.

What Are The Main Stages of Viral Reproduction in Host Cells?

This process can be broken down into five main stages:

First of all - adsorption - the virus binds to the outside of the host cell.

Secondly - penetration - the virus introduces its genetic material (either RNA or DNA) into the cell.

Thirdly - reproduction of viral genetic material occurs - this takes place using the cell's own machinery (i.e. enzymes) and occurs in the cell's nucleus.

Fourthly - assembly - viral protein components and enzymes are produced at the host cell's ribosomes using mRNA made from viral genes. These components then begin to be assembled.

Finally - maturation and release - new viral particles, copies of the original infecting virus, are made - with, as originally, a piece of RNA or DNA inside a protein coat. These viral particles leave the cell, usually destroying the cell in the process. Of course these viruses can then infect other cells.

So, when a virus infects a cell, it uses the cell's organelles such as ribosomes and much of the cellular machinery to replicate.

How Do Viruses Increase in Number So Rapidly?

Whereas the reproduction of living cells uses either mitosis or meiosis to create two or four daughter cells respectively; viral replication produces many thousands of copies of the virus that leave the host cell to infect other cells in the organism.

This means that viruses can multiply in number very rapidly.

Why Do Viruses Infect Only A Specific Type of Cell?

Viruses typically can only infect a limited number of types of host cells (known as the virus' host range).

The reason for this is that there are certain proteins of a certain shape on the virus particle which must fit certain-shaped receptor molecules only found on the surface of the cell membrane of a particular host cell type.

What Is A Retrovirus?

Retroviruses are viruses that have RNA as their genetic material.

What Is Different About The Way In Which Retroviruses Replicate?

Retroviruses are different in at least three respects from usual viruses.

Firstly, retroviruses replicate using reverse transcription of their RNA within the host cell that they infect. Reverse transcription of RNA produces DNA as follows: a reverse transcriptase enzyme, originally contained within the virus, makes the complementary DNA strand from the viral genome RNA template. This DNA then gets integrated into the host cell genome and instructs the cell to make more copies of the virus.

Secondly, retroviruses are unusual because they enter their host cell intact. Once inside the cell, their protein coat is removed, releasing the single stranded piece of RNA that is the virus' genetic material and serves as the template for the synthesis of new viral RNA genomes. Protein synthesis from these genomes results in development of new viral particles.

Thirdly, retroviruses are unusual because they are released slowly from the host cell by a slower process called budding which does not cause lysis (splitting up) or death of the host cell, as other viruses do.

What Diseases Are Caused By Retroviruses?

Retroviruses are responsible for causing some forms of leukaemia) in humans, and AIDs (HIV).

Chapter Three

Diseases Caused By Viruses

HIV/AIDS

What Is AIDS?

AIDS is a disease that is characterised by the sufferer having an immune system is too weak to fight off infections.

What Is The Difference Between HIV and AIDS?

HIV is the virus that causes AIDS to develop.

Someone may have been infected with the HIV virus and never develop AIDS; because the virus remains dormant in their body.

In fact; the average time between getting the HIV infection and getting AIDS is around ten years; the length of time is thought to depend on many factors including general health and lifestyle.

Medications can now increase the time it takes AIDS to develop and may prevent it from developing at all.

What Are The Symptoms of AIDS?

They vary between patients; but in general; due to the weakening of the immune system; the patient suffers from different infections; causing a subsequent loss of physical condition and poor general health.

Frequently patients eventually get one of the; what is called 'opportunistic'; infections; which infect the patient thanks to the 'opportunity' of their weakened immune system; e.g. tuberculosis; pneumonia, thrush and herpes; plus forms of cancer such as Kaposi's sarcoma.

Ultimately, these infections can lead to death.

Which Virus Causes AIDS?

The human retrovirus HIV (Human Immunodeficiency Virus) causes Acquired Immune Deficiency Syndrome (AIDS).

How Many People Suffer From AIDS?

It is estimated that in 2013 around thirty million people worldwide (that is about half the population of Britain or about the population of the state of California) were infected with the HIV virus and that about two million people a year died of AIDS; with about a quarter of a million of deaths being children.

How Does Being Infected With The HIV Virus, Lead To Developing AIDS?

The HIV virus destroys white blood cells called T cells, which are vital for the function of the human immune system. As HIV attacks these cells, the person infected with the virus is less able to fight off infection and disease, ultimately resulting in the development of AIDS.

Most people who are infected with HIV can carry the virus for years before developing any serious symptoms. But over time, HIV levels increase in the blood and number of T cells decrease. People who are not infected with HIV and are healthy usually have about one thousand T cells per mm^3 of blood, while people with AIDS have been known to have less than fifty T cells in their entire body (as little as five per cent of the number in healthy people).

Antiretroviral medications can help reduce the rate of increase of HIV levels and therefore help to maintain T cell numbers; which can slow down or prevent the development of AIDS.

How Is AIDS Transmitted Between People?

The HIV virus can be found in bodily fluids: the blood, semen, urine, faeces, milk and vaginal lubrication; so it can be transmitted between people when the bodily fluids of an infected person enter the body of another person; the virus infects the other person by moving through a small cut, blister, or break in their skin or within their body parts.

A common means of transmission is during unprotected sexual intercourse (sex without the use of a condom/prophylactic/rubber); another is through the sharing of needles by addicts.

How Can AIDS Be Treated?

There is no cure for AIDS - although drugs can help stop HIV infection from causing AIDS to develop.

There are medications that are now available to treat HIV infection.

Some prevent the virus from copying itself inside the immune system cells. Others stop the virus maturing and leaving the cells. None of these drugs cures HIV infection (as the virus remains in the patient's body) or AIDS, although they can delay the appearance of AIDS, sometimes permanently.

These drugs are expensive and many people around the world with HIV cannot afford these drugs.

How Can AIDS Be Prevented?

To prevent transmission the person must avoid the possibility contact with the bodily fluids of any infected person that could lead to the virus entering their body through a small cut, blister or broken skin.

In practical terms this means not having sexual intercourse without using a male or female condom/prophylactic/rubber (it is thought to be extremely unlikely for the HIV virus to be transmitted by kissing an infected person). Drug users should not share needles. People such as health workers and carers who deal with possibly infected people who are bleeding (or clean up their urine or faeces) should make sure they wear latex gloves and covering up any cuts they might have.

POLIO

What Is Polio?

Polio, or poliomyelitis, is a highly contagious viral infection that can lead to paralysis, breathing problems, or even death.

Which Virus Causes Polio?

Polio is caused by the poliovirus.

How Does The Poliovirus Cause Polio?

The poliovirus is a highly contagious virus specific to humans. It invades the nervous system, and can cause total paralysis in a matter of hours

How Does Someone Become Infected With Polio?

The virus can be found in the faeces of an infected person. The virus can spread to other people via water or food contaminated with faeces as well as (less commonly) through direct contact with the saliva or respiratory droplets of an infected person.

What Kind Of Person Catches Polio?

Polio mainly affects children under five years of age.

How Many People Have Polio?

In 1988, the forty-first World Health Assembly adopted a resolution for the worldwide eradication of polio mainly through vaccination and surveillance. This has proved to be remarkably successful. In 1998 there were an estimated three hundred and fifty thousand cases worldwide - but the global effort to eradicate it has reduced this by about 99% to just a few hundred cases in 2013.

In 2014, only three hundred and fifty polio cases have been reported worldwide; from the following countries: Afghanistan, Cameroon, Equatorial Guinea, Ethiopia, Iraq, Nigeria, Pakistan, and Syria.

What are the symptoms of polio?

There are different forms of polio. The more serious forms can give rise to symptoms such as paralysis and may cause death but many people who are infected with polio do not become ill at all and have no symptoms.

The symptoms that do appear depend on the type of polio the person has contracted.

Non-paralytic polio causes flu-like symptoms for a week or two, often along with back and neck pain and muscle stiffness.

Paralytic polio starts in a similar way to non-paralytic but continues with severe muscle pain and spasms, and paralysis of limbs; particularly on one side of the body. One in every two hundred infections leads to irreversible paralysis; of these around one in every fifteen will die due to not being able to use their breathing muscles. This means that around one in every three thousand cases will end in death.

How Is Polio Diagnosed?

Polio can be diagnosed from characteristic symptoms such a lasting neck and back stiffness, abnormal reflexes, and trouble with swallowing and breathing. Laboratory tests for the presence of the poliovirus can be made from throat swabs, stool samples, or cerebrospinal fluid.

How Is Polio Treated?

There is no cure. Treatments are available that can help manage symptoms, and prevent worsening of the condition or complications e.g. antibiotics may be given for additional infections; also pain killers may help the patient be more comfortable; ventilators can help to ease breathing; muscle tone can be improved with exercise and physiotherapy, and a good diet can improve the patient's overall condition.

How can polio be prevented?

Polio is best prevented through a vaccine, given multiple times, which can protect a child for life. Effective sewage disposal and clean water also help to prevent it from spreading.

INFLUENZA (FLU)

What is influenza (also called flu)?

Flu is a respiratory illness caused by influenza viruses that infect the nose, throat, and lungs. It can vary from a mild to severe illness, in some cases causing death.

How Can Flu Be Prevented?

The best way to prevent the flu is by getting vaccinated against it every year.

How Is Flu Spread (Transmitted) From Person to Person?

The influenza virus is contagious.

Flu viruses mainly spread by droplet infection. When people who have flu, cough, sneeze, or talk; droplets of mucus from their nose, throat, mouth or lungs can float in the air and be breathed in by other people who are near. The body then contracts the flu virus through mucus membranes in the nose, mouth or eyes.

Adults are thought to be able to infect others from one day before the flu symptoms develop until up to five days afterwards.

It is also possible that flu can be contracted by someone touching a surface or object that has flu viruses on it and then transferring them to their mouth, eyes and nose.

Can Flu Be A Serious Disease?

The worldwide flu epidemic of 1918 is estimated to have killed more than twenty million people, including half a million Americans.

As recently as 1998-99; around sixty five thousand people are thought to have died from flu in America.

Flu is unpredictable and how severe it is can vary widely depending on who contracts it and what flu viruses are spreading. People who are at greater risk of serious complications from having flu include: people over 65 years old, pregnant women, children and people with asthma, diabetes, and heart disease.

Worldwide; between 5% and 10% of adults and 20%–30% in children get flu every year. Hospitalization and death usually occur in mainly the high-risk groups such as the very young, elderly or chronically ill. Worldwide, it is thought that there are up to five million cases serious cases of flu which ultimately result in up to half a million deaths.

It is estimated that between 1976 and 2006, as many as fifty thousand people died from flu or complications caused by flu. These complications of flu often include bacterial pneumonia, along with worsening of the symptoms of heart failure, asthma, or diabetes. It is believed that around 90% of deaths from flu are due to older patients who develop pneumonia as a complication.

What Are The Symptoms Of Flu?

Flu is characterised by a sudden arrival of a number of symptoms (some or all of which may be present) such as: fever, headache, dry cough, sore throat, muscle aches, chills and sweats, loss of appetite, aching joints, runny nose and excessive fatigue.

YELLOW FEVER

What Is Yellow Fever?

Yellow fever is a viral infection found in tropical areas. It is transmitted by mosquitoes.

Yellow fever is passed to humans by bites from infected mosquitoes which tend to bite during daylight hours. (This is different to the mosquitoes which carry malaria, which tend to bite from dusk to dawn.) Yellow fever occurs in certain countries of tropical Africa and South America. Yellow fever is not transmitted directly from person to person.

Yellow fever can be a serious disease.

What Are The Symptoms of Yellow Fever?

For some people it causes relatively mild symptoms are similar to flu; e.g. headache, aching joints, fever which improves completely.

However, for many people it can be more serious; causing high fever, vomiting, and jaundice (which is what gives the illness its name - people with jaundice have a yellow skin colour).

In really serious cases there can be internal bleeding; kidney problems and meningitis; which can all be fatal.

There is no cure for yellow fever and the mortality rate in epidemics can high; at over fifty percent.

What Causes Yellow Fever?

Yellow fever is caused by a virus which belongs to the Flaviviridae family. The virus is transferred from an infected person to a non-infected person via a mosquito bite that introduces the virus into the bloodstream via the saliva of the mosquito.

The virus can reproduce itself in a number of different types of body cells – such as liver, kidney and blood vessels. In serious cases, these cells (and thereby; the organs they make up) become damaged.

Certain cells of the immune system are affected and release large quantities of signalling substances. These substances are the cause of the normal disease symptoms, such as muscular pain and fever that are also observed in influenza.

How Is Yellow Fever Passed On?

Yellow fever has three cycles of infection - a jungle cycle; where it infects first of all, monkeys and is then transferred by mosquito bites to humans - and an urban cycle where it infects only humans. There is also an intermediate cycle.

The virus is transmitted by a couple of species of mosquito.

In its original jungle cycle, the mosquito bites an infected monkey; the mosquito becomes infected with the virus; which leads to the virus accumulating in the mosquito's salivary glands. When the infected mosquito bites another monkey, the monkey becomes infected with the virus. So the mosquito transmits infection between monkeys. A person travelling through the jungle who is bitten by an infected mosquito will become infected with the virus. When this person returns to urban areas, the urban cycle begins.

Urban cycles start when an infected traveller returns from the jungle. A mosquito bites the traveller, so the mosquito becomes infected and passes the virus on by biting other people.

There is also an intermediate cycle often found in Africa where people in villages near the jungle and monkeys in the jungle are both bitten by a transmitting mosquito that breeds in both areas.

Where Does Yellow Fever Occur?

This viral disease regularly occurs in tropical areas of Africa and South America and on some Caribbean islands; in a total of more than thirty countries; making more than five hundred million people at risk of contracting it.

There is a risk of it spreading to Asia (where there are mosquitoes that could spread it) which Asian countries try to prevent by insisting on a vaccination certificate for people who travel from tropical, affected areas.

How Many People Become Infected With Yellow Fever?

Each year there are estimated to be about two hundred thousand cases of yellow fever worldwide, of which around thirty thousand result in death, although this is probably an underestimate as it based on generally quite unreliable records. Most of the deaths occur in Africa.

What Are The Symptoms Of Yellow Fever?

The incubation period from first being infected with the virus to developing the symptoms of yellow fever is between a few days and about two weeks.

Initial symptoms include muscle aches, fever, nausea, and vomiting. In many patients there will be improvement in symptoms and gradual recovery occurring three to four days after the onset of symptoms.

But within a day of an apparent recovery, up to a quarter of patients develop a more serious form of the illness and of these up to half may die. Anaemia (lack of red blood cells) develops, as well as hepatitis and jaundice (which is characterised by yellow skin and eyes - from which yellow fever gets its name). The kidneys may fail and bleeding from the mouth, nose and stomach may occur along with blood contained in vomit and faeces. Unfortunately most patients who show this bleeding will die within a short time period.

How Can Yellow Fever Be Prevented?

The best method of prevention is by vaccination against yellow fever. This gives protection against the disease after about ten days from vaccination and lasts about ten years.

As the disease is spread through mosquito bites it is possible that prevention of mosquito bites can help in avoiding yellow fever. This method of prevention though is not reliable and vaccination should always be used where possible.

However in some poor countries vaccination is not available and bite avoidance may then be the only method of protection.

How Is Yellow Fever Diagnosed?

The disease is usually difficult to tell apart from other illnesses in its early stages. A blood sample that detects specific yellow fever virus antibodies can give a definite diagnosis.

How Is Yellow Fever Treated?

No medications are effective in curing yellow fever; it is only possible to alleviate the symptoms, through the use of painkillers and anti-febrile medication to bring down the patient's temperature. Dehydration can be treated with intravenous saline solution.

Serious cases of yellow fever require hospitalisation of the patient.

Chapter Four

How Viruses Can Benefit Humans

How Might Viruses Be Useful To Humans?

At first glance viruses seem like a totally destructive things; killing cells or causing disease; not doing any good at all.

From the viewpoint of biological ecosystems, a virus is purely a parasite, and purely destructive. It destroys cells, plant and animal or bacterial, simply in order to replicate itself (using its RNA, ribonucleic acid, or DNA, deoxyribonucleic acid), and repeats the cycle as long as it can find a host cell.

Usually this is true, but for biotechnology applications, viruses can be used to carry useful genes from one cell to another. Genetic engineers use viruses for this purpose; using viruses as what are called 'vectors' in gene therapy.

With many vaccines, an attenuated virus is introduced to provide the host (patient) immunity against possible future invasion by a healthy virus.

How Are Viruses Used in Biological Studies?

Viruses are used in molecular and cellular biology studies. They are used to manipulate and investigate the functions of cells.

Viruses have been used extensively in genetics research and understanding of the genes and DNA replication, transcription, RNA formation, translation, protein formation and basics of immunology.

How Are Viruses Used In Medicine?

Viruses have a number of uses in medicine. They are used as vectors to carry the chemicals for treatment of a disease to various target cells.

There has been research into their use as vectors to carry healthy copies of genes to cells, as part of gene therapy for genetic diseases such as cystic fibrosis. Gene therapy is the introduction of new, functioning genes into the cells of a patient suffering from an inherited genetic disease. This new gene should allow the cell to perform correctly its functions and corrects defective or non-operational inherited genes within those cells.

Viruses used in this way are like vehicles carrying 'good' genes into human cells; where these genes may be used (instead of the 'unhealthy' inherited form of the gene) successfully by the cell.

Viruses have been used as gene vectors in the successful treatment of a genetic disease called

Severe Combined Immunodeficiency (SCID). However, this therapy has the risk of causing mutation in existing healthy genes and also of over-expression of the introduced gene, e.g. in the SCID trial sadly, some patients developed leukaemia as a result of this kind of mutation.

How Are Viruses Used In Bacteriophage Therapy?

Bacteriophages are highly adapted viruses that specifically target, infect, and destroy bacteria.

Bacteriophages are believed to be the most numerous types of viruses accounting for the majority of the viruses present on Earth. These are basic tools in molecular biology. They have been researched for their use in therapy.

The emergence bacteria resistant most (sometimes all) antibiotics has become a critical problem in modern medicine. The development of alternative antibacterial treatments is, as a result, one of the highest priorities of modern medicine.

It has been suggested that bacterial infections could be treated by the use of bacteriophages. Although not so far used in the West, this type of bacteriophage therapy has been successfully used to treat bacterial infection in parts of the former Soviet Union and Eastern Europe. Research continues into its possible future use in the West.

How Are Viruses Used In The Genetic Engineering of Crop Plants?

Modification and genetic engineering methods can be used to make modified genomes that can be carried into plants cells by viruses acting as vectors.

This method can lead to more productive transgenic crop plants.

How Are Viruses Used To Make Vaccines?

Vaccines use a destroyed or weakened form of the pathogen that causes the illness (in this case a virus) to stimulate the production of antibodies against it by white blood cells in the body - this should result in the person being immune to any future exposure to this particular virus.

In fact, viruses were used by Edward Jenner in the first ever vaccine.

Nowadays, the commonly given vaccines against the viral diseases: polio, measles, chicken pox; use weakened viruses. So a vaccination against measles would be in the form of weakened measles viruses.

This use of vaccination has been one of the great public health advances since the middle of the twentieth century. Indeed a number of diseases have been virtually eradicated through vaccination; and the virus that causes smallpox was completely eradicated by a worldwide vaccination program; leading to the last case being reported in 1977.

Viral vaccines can even prevent certain kinds of cancer. Vaccines for hepatitis B and those for human papillomavirus can be given to help prevent respectively liver and cervical cancer.

What Is Virus-directed enzyme prodrug therapy (VDEPT)?

This is a very useful new type of cancer chemotherapy. It enables the cancer drugs to act specifically in the tumour cells only; reducing the side effects of toxic effects on other body cells.

An enzyme is inserted into the tumour cells through the introduction to them of a gene; by using a virus that carries the gene. The enzyme activates an inactive form of a cytotoxic drug that can therefore be administered systemically because the active, cytotoxic form of the drug is only produced in the tumour cell - where the relevant enzyme is present and active.

This therapy is currently being used to treat HIV infection. An adenovirus with the thymidine kinase (TK) enzyme of the herpes simplex virus is combined with systemic administration of a medication called ganciclovir. This is changed by the TK enzyme into the active form only in cells where the TK enzyme is present. This is used to target action on the infected white blood cells called T helper cells in HIV treatment.

How Are Viruses Used To Control Agricultural Pests?

Viruses can also be used to control damaging pests. Traditionally chemical pesticides have been used to control pests in agriculture, but these can harm the environment by getting into the food chain, and can harm human health. Viral control has also been recognized as inherently less toxic than conventional pesticides.

Viruses used for pest control commonly cause disease of the target species. Viruses are used for the control of multiple species of pest insects and also for rabbits (Myxomatosis).

Their disadvantages include limited range of action, slow effects compared to chemical agents, high costs of initial treatment, low environmental stability, particularly in sunlight etc.

Chapter 5

The Evolution of Viruses

Are There Any Fossil Records of Viruses?

Viruses don't leave a fossil record because they are too tiny and insubstantial, so it is difficult to determine the evolutionary origin of viruses.

However traces of viral DNA has been found integrated into the DNA of some really old fossils remains, demonstrating that viruses were present a long way back in evolutionary history.

How Did Viruses Originate?

There are three hypotheses as to how viruses originated.

One hypothesis is that they evolved before other forms of life around four billion years ago, at the beginning of life on Earth. This is called the virus first hypothesis and presumes that viruses either came into existence before cells or were created at the same time as the first cells. However, a major flaw in this hypothesis is that viruses need host cells in order to reproduce and so couldn't logically exist before cells evolved. In this case they may have come into existence at the same time, or just after cells did.

Another hypothesis is that they evolved from components of cells that gained the ability to replicate and spread to other cells. This is called the escape hypothesis and assumes that viruses are actually pieces of RNA or DNA that escaped from cells and then began invading other cells. This hypothesis has the weakness that it cannot explain the existence of viral structures such as capsules that surround the virus.

The third hypothesis is that viruses evolved from another type of inter-cellular parasite. This is called the reduction hypothesis and states that viruses were once cells that were parasites of larger cells. Gradually these parasitic cells lost functions and organelles (hence 'reduction'); eventually evolving into viruses are they are today. This neatly explains much of why host cells are needed for viruses to reproduce, but it cannot account for the fact that small parasites do not resemble viruses at all.

How Can Viruses Evolve?

Evolution is based on the action of natural selection on variation.

Viruses are able to evolve because a number of features in the biology of viruses cause allow them to be able to generate a lot of variation rapidly. Firstly, viruses have high mutation rates because copying viral DNA doesn't involved the host cell's DNA checking mechanisms and this creates a large amount of viral genetic variation. In addition recombination and re-assortment can also produce more genetic variation.

Secondly, viruses produce numerous offspring within a short time allowing fast adaptation and the spread of advantageous characteristics.

These mutations can cause the viruses to quickly change over a short period of time driving viral evolution to be done at very high speeds.

www.ingramcontent.com/pod-product-compliance
Lightning Source LLC
Chambersburg PA
CBHW070924180526
45168CB00005B/2144